Leaping Learners
Education, LLC

For more information and resources, visit us at:
www.leapinglearnersed.com

Every attempt has been made to credit each photo. Please contact us if there has been an error and we will resolve the issue.

Photo Credits

Cover, © Calek/stock.adobe.com, Fish diver graphics © Alexey Bannykh/stock.adobe.com; pg. 9-10 ©ead72/stock.adobe.com; pg 11-12 ©uwimages/stock.adobe.com; pg 13 ©nicolaprimola/stock.adobe.com; ©laurentdambies/stock.adobe.com; pg 14 ©uwimages/stock.adobe.com; pg 15 ©cherylub/stock.adobe.com, ©mirko Rosenau/stock.adobe.com; pg 16©Ocean-Image/stock.adobe.com

All designs by Sean Bulger

All other pictures by Sean Bulger or royalty-free from Pixabay.com

Special thanks to our reading level consultant

FRANCINE DEL VECCHIO, Ed.D.

Director of the Literacy Center

Professor of Education

Caldwell University

ISBN

978-1-948569-05-7

Dear Parents and Guardians,

Thank you for purchasing a *Fay Learns About* series book! After teaching students from kindergarten to second grade for over seven years, I became frustrated at the lack of engaging books that my students could read independently. To help my students engage with nonfiction topics, my wife and I decided to write non-fiction books for children. We hope to inspire young children to learn about the natural world.

Here at Leaping Learners Education, LLC, we have three main goals:

1. Spark young readers' curiosity about the natural world
2. Develop critical independent reading skills at an early age
3. Develop reading comprehension skills before and after reading

We hope that your child enjoys learning with Fay. If you or your children are interested in a topic that we have not written about yet, send us an email with your topic and maybe your book will be next!

Thank you,

Sean Bulger Ed.M

www.leapinglearnersed.com

Reading Suggestions:

Before reading this book, encourage your children to do a "picture walk," where they skim through the book and look at the pictures to help them think about what they already know about the topic. Thinking about what they already know helps children get excited about learning more facts and begin reading with some confidence.

Preview any new vocabulary words with your child. Key vocabulary words are found on the last few pages of the book. Have your children use each new phrase in their own words to see if they understand the definition.

After previewing the book, encourage your child to read the book independently more than once. After they have read it, ask them specific questions related to the information in the book. Encourage them to go back and reread the relevant section in the book to retrieve the answer in case they forgot the facts.

Finally, see if your child can complete the reading comprehension exercises at the end of the book to strengthen their understanding of the topic!

This book is best for ages 7-8
but. . .
Please be mindful that reading levels are a guide and vary depending on a child's skills and needs.

Fay Learns About . . . Clownfish

Written by Sean and Anicia Bulger

Table of Contents

Hi! My name is Fay.
I love to discover
and learn new
things. In this book,
we will learn about
clownfish. Let's go!

Introduction

Clownfish are saltwater fish that live in the ocean.

Body

Clownfish have two eyes and a mouth.

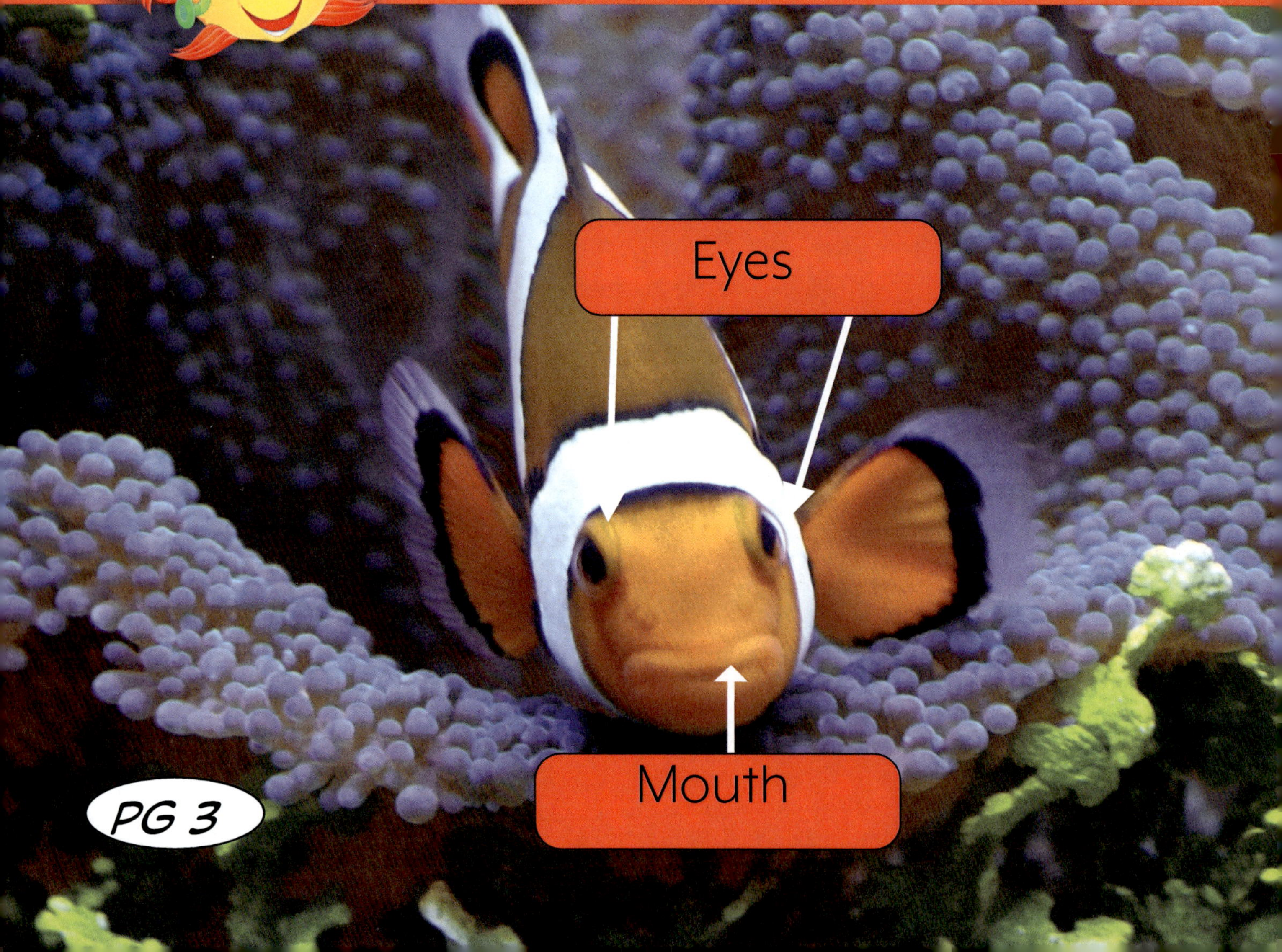

Clownfish have eight fins. Two on top, two on their sides, three on the bottom, and one tail fin.

Habitat

Clownfish live in warm water around coral reefs. They make their home in an animal called a **sea anemone**.

Tricky Word
anemone
say.....
a-nem-o-ne
Anemone

Sea anemones are **toxic** to other animals, but clownfish are not hurt by them because they slowly rub their bodies against them and become **immune** to the toxins.

Tricky Word
IMMUNE
chunk it out
IM - MUNE

Babies

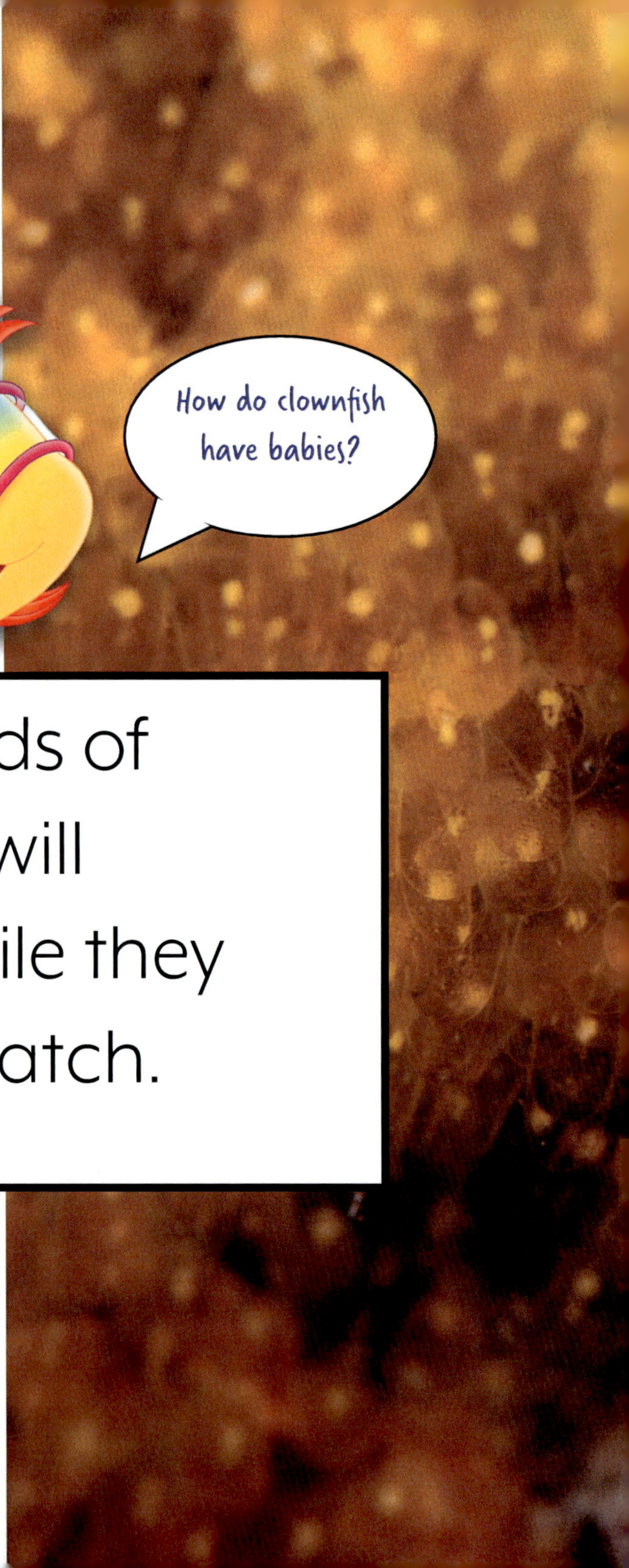

Clownfish lay hundreds of eggs at a time. They will protect their eggs while they wait for the eggs to hatch.

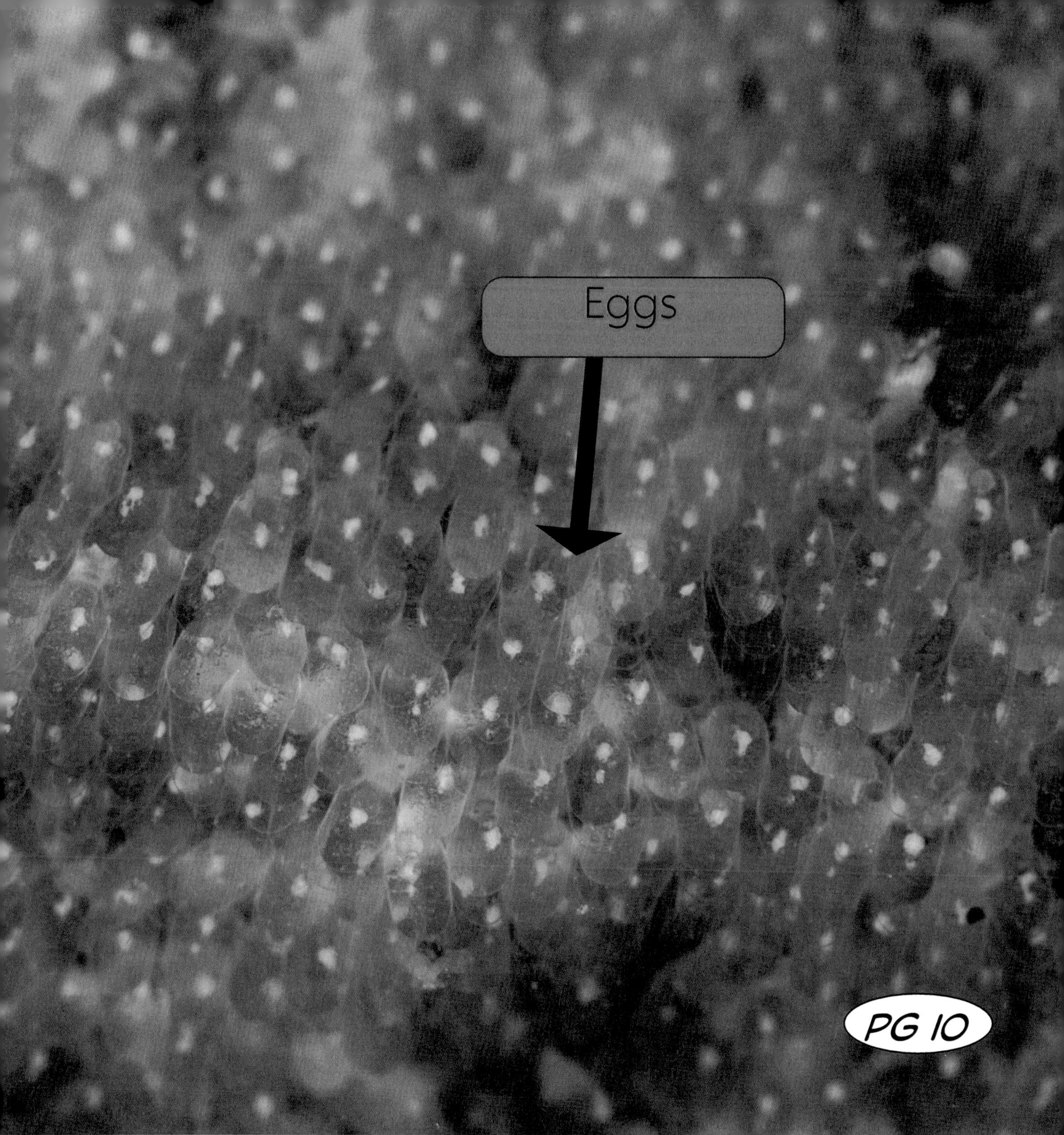
Eggs

Baby clownfish are small. They will live with their parents until they are adults.

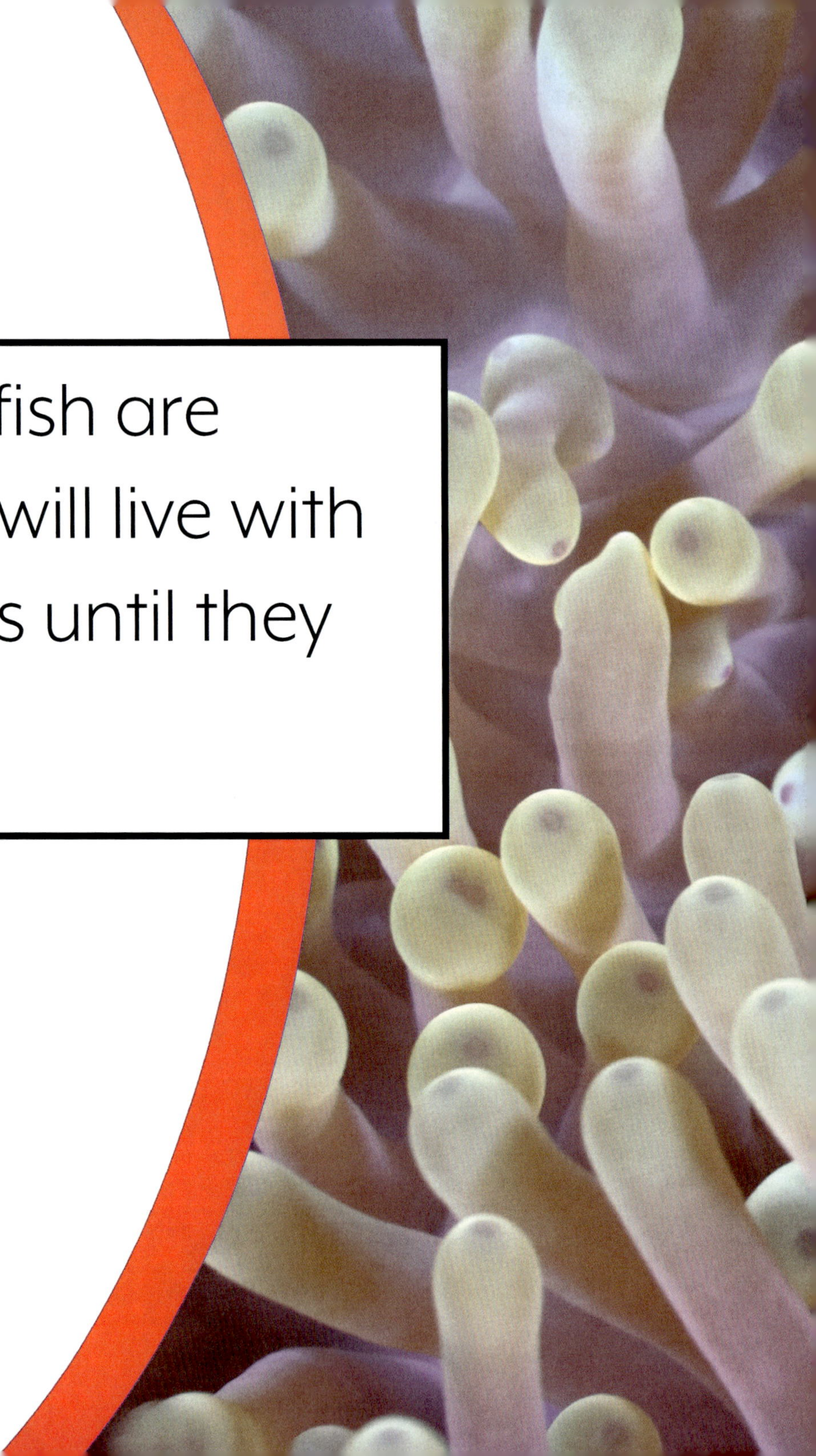

Diet

Clownfish eat dead parts of the sea anemone they live in. They also eat **plankton** and **algae**.

Plankton

Algae

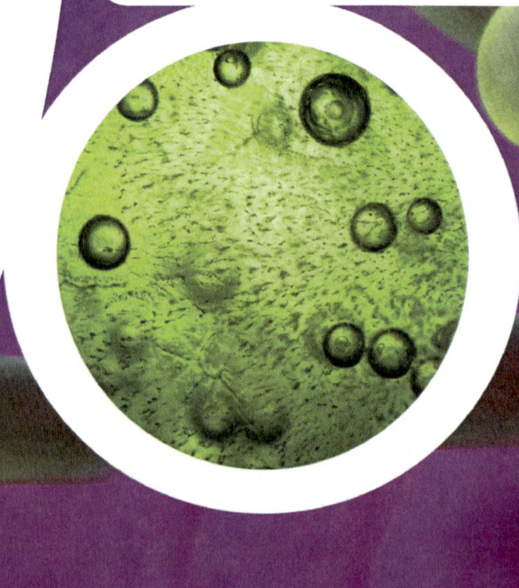

Predators

Clownfish are hunted and eaten by eels, large fish, and sharks.

Sharks

Fish

Eel

All clownfish are born male! Some change to female when they get older.

The character *Nemo* was a clownfish!

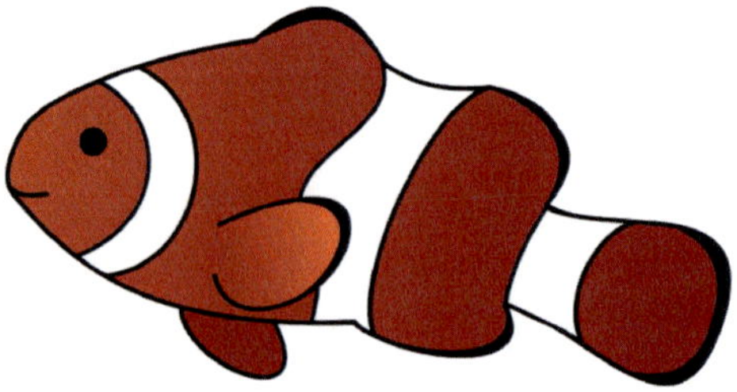

1,000!

Clownfish can lay up to 1,000 eggs at a time!

Glossary

A glossary tells the reader the meaning of important words.

Sea anemone – A colorful soft-bodied ocean animal that eats other fish

Toxic – A poison that kills animals or makes them sick

Immune – Not affected by poison or disease

Algae – Plant-like organism that gets its energy from the sun

Plankton – Tiny living organisms that float and live in the ocean

Draw a picture of a clownfish.

Choose 3 words from the glossary and write a sentence for each one.

1. ______________________________

____

____

2. ______________________________

____

____

3. ______________________________

____

____

Quiz

1. How many fins does a clownfish have?

a. 2

b. 6

c. 8

2. What is one thing that clownfish eat?

a. Snail

b. Plankton

c. Eel

3. How do clownfish stay safe?

a. They live in toxic sea anemones.

b. They shoot poison.

c. They have a shell.

Common core standards:

RI. 2.1 - Questions 1, 2, 3

4. About how many eggs can one clownfish lay?

a. one

b. ten

c. hundreds

5. In which section do you learn about where clownfish live?

a. Body

b. Habitat

c. Diet

6. How are all clownfish born?

a. Female

b. Male

Common core standards:
RI. 2. 1 - Questions 4, 6
RI. 2. 2 - Question 5

Answer the following questions using information from the text:

1. Describe what a clownfish looks like.

____

____

____

____

____

2. Why can clownfish live in a toxic sea anemone?

____

____

____

____

____

____

3. What do clownfish eat?

4. Why would a clownfish avoid eels?

Common core standards:
RI. 2. 1 - Question 1, 3
RI. 2. 8 - Questions 2, 4

Want to learn about rainforest animals? Check out the "Matt Learns About..." series!

Great for emerging readers ages 6-8

Want to learn about colors? Check out the "Clayton Teaches you About..." series!

Great for early readers ages 4-6

Want to learn about Farm Animals? Check out the "Katie Teaches You About..." Series!

Great for early readers ages 4-6